Seasons in a Year

FALL

AMY CULLIFORD

A Crabtree Roots Book

CRABTREE
Publishing Company
www.crabtreebooks.com

School-to-Home Support for Caregivers and Teachers

This book helps children grow by letting them practice reading. Here are a few guiding questions to help the reader with building his or her comprehension skills. Possible answers appear here in red.

Before Reading:

• What do I think this book is about?
  • *This book is about a season called fall.*
  • *This book is about things you can see in fall.*

• What do I want to learn about this topic?
  • *I want to learn what fall looks like.*
  • *I want to learn what colors the leaves turn in fall.*

During Reading:

• I wonder why...
  • *I wonder why leaves turn colors.*
  • *I wonder why it gets cool in the fall.*

• What have I learned so far?
  • *I have learned what fall looks like.*
  • *I have learned that leaves turn red and yellow in fall.*

After Reading:

• What details did I learn about this topic?
  • *I have learned that leaves can change into different colors.*
  • *I have learned that we go back to school in the fall.*

• Read the book again and look for the vocabulary words.
  • *I see the word **frost** on page 10 and the word **leaves** on page 4. The other vocabulary words are found on page 14.*

What do you see in **fall**?

I see red **leaves**.

I see brown leaves.

I see a **pumpkin**.

I see **frost**.

I see my **school**.

Word List
Sight Words

a	in	you
do	my	what
I	see	

Words to Know

fall

frost

leaves

pumpkin

school

25 Words

What do you see in **fall**?

I see red **leaves**.

I see brown leaves.

I see a **pumpkin**.

I see **frost**.

I see my **school**.

Seasons in a Year

FALL

Written by: Amy Culliford
Designed by: Rhea Wallace
Series Development: James Earley
Proofreader: Kathy Middleton
Educational Consultant: Christina Lemke M.Ed.
Photographs:
Shutterstock: Lilkar: cover; rdonar: p. 1; Jenny Sturm:
   p. 3, 14; Ulrich Mueller: p. 5, 14; Dino Osmic: p. 7;
   Leena Robinson: p. 9, 14; Maya Kruchankova: p.
   10-11, 14; James R. Martin: p. 13, 14

Library and Archives Canada Cataloguing in Publication
Title: Fall / Amy Culliford.
Names: Culliford, Amy, 1992- author.
Description: Series statement: Seasons in a year | "A Crabtree roots book".
Identifiers: Canadiana (print) 20200387073 |
   Canadiana (ebook) 2020038709X |
   ISBN 9781427134721 (hardcover) |
   ISBN 9781427132697 (softcover) |
   ISBN 9781427132734 (HTML) |
   ISBN 9781427133113 (read-along ebook)
Subjects: LCSH: Autumn—Juvenile literature.
Classification: LCC QB637.7 .C85 2021 | DDC j508.2

Library of Congress Cataloging-in-Publication Data
Names: Culliford, Amy, 1992- author.
Title: Fall / Amy Culliford.
Description: New York : Crabtree Publishing Company, 2021. | Series:
   Seasons in a year : a Crabtree roots book | Audience: Ages 4-6 |
   Audience: Grades K-1 | Summary: "Early readers are introduced to
   the fall season. Colorful pictures and simple sentences make fall
   come alive"-- Provided by publisher.
Identifiers: LCCN 2020049770 (print) |
   LCCN 2020049771 (ebook) |
   ISBN 9781427134721 (hardcover) |
   ISBN 9781427132697 (paperback) |
   ISBN 9781427132734 (ebook) |
   ISBN 9781427133113 (epub)
Subjects: LCSH: Autumn--Juvenile literature.
Classification: LCC QB637.7 .C85 2021 (print) | LCC QB637.7 (ebook) |
   DDC 508.2--dc23
LC record available at https://lccn.loc.gov/2020049770
LC ebook record available at https://lccn.loc.gov/2020049771

Crabtree Publishing Company

www.crabtreebooks.com   1-800-387-7650

Printed in the U.S.A./022021/CG20201130

Published in the United States
Crabtree Publishing
347 Fifth Avenue, Suite 1402-145
New York, NY, 10016

Published in Canada
Crabtree Publishing
616 Welland Ave.
St. Catharines, Ontario L2M 5V6